Nutrition
EDUCATION
FOR KIDS

Health Science Series

Katherine Johnson

authorHOUSE®

AuthorHouse™
1663 Liberty Drive
Bloomington, IN 47403
www.authorhouse.com
Phone: 1-800-839-8640

Published by AuthorHouse 6/6/2013

ISBN: 978-1-4817-0099-3 (sc)
ISBN: 978-1-4817-0100-6 (e)

Library of Congress Control Number: 2012924011

Nutritional Science aims to understand the relationship between diet, health and disease. It explores what food components increase the risk factor for chronic diseases, while posing questions such as, which diet work best for optimal health. Focus areas of study in nutritional sciences specialization includes: biochemical and molecular aspects of nutrient metabolism; the impact of diet on health of individuals and populations; and nutritional - related policy and public health issues. A degree in nutrition accords a wide array of career choices, including public health nutritionist, nutrition educator, or research scientist in biotechnology, government or medicine.

NUTRITION

Did you know nutrition is the process of eating a healthy diet necessary for growth and wellness?

A healthy diet consists of consuming vital nutrients our body needs.

NUTRIENTS

Food is made up of all kinds of nutrients. These nutrients are all the important substances our body needs for growth and maintenance of life.

Macronutrients are substances our body needs in large amounts. Micronutrients are substances our body needs, but only in small amounts. We get both of these kinds of nutrients from the food we eat every day.

MACRONUTRIENTS

Water, protein, carbohydrates, and fats are four kinds of macronutrients.

Each of these nutrients are essential for growth and development of our body.

Macro means large. This will help you remember that we need macronutrients in large amounts.

WATER

Water is the most important macronutrient.

It helps dissolve other nutrients from the foods we eat and transport them to all the tissues in our body. Every day we need water in larger amounts because our body is mostly water. And water is part of every cell in our body.

PROTEIN

Protein is a macronutrient found in food.

We need protein to grow tall and build muscles. Beef, poultry, fish, eggs, dairy products, seeds, legumes, and lentils are all examples of foods high in protein. They come from both animals and plants.

Foods from animals are complete proteins. Foods from plants are incomplete proteins.

BEEF

Beef is meat from cattle.

Cattle can be a cow or a bull.

Meat from cattle is a complete protein.

POULTRY

Poultry is meat from chickens and turkeys.

Poultry is a complete protein.

FISH

Fish are animals that live in oceans, rivers, and lakes.

There are many different kinds of fish to eat.

Fish are also a complete protein.

EGGS

Chickens, ducks, and quails all lay eggs that are eaten by humans.

Eggs are a complete protein.

DAIRY PRODUCTS

Dairy products are foods made from the milk of cows or goats.

Cheese, yogurt, and ice cream are all dairy products.

SEEDS

Seeds are tiny plants enclosed in shells.

The hard shell is called a seed coat.

Seeds are an incomplete protein.

LEGUMES

Legumes [leg-ume] are dry beans and peas.

These foods are an incomplete source of protein.

LENTILS

There are many different kinds of lentils.

A lentil is a plant of the legume family.

They are dried lens-shaped seeds.

Lentils are also an incomplete protein.

COMPLETE PROTEINS

Complete protein foods contain all nine essential amino acids our body needs for growth,

health, and maintenance of life. Animal products are complete protein foods.

Meat, fish, poultry, cheese, eggs, yogurt, and milk are all complete proteins.

INCOMPLETE PROTEINS

An incomplete protein lacks one or more of the nine essential amino acids.

All protein from plants is incomplete by itself. That means that grains, seeds, legumes, and peas lack one or more amino acids our body needs. But when we combine foods from two or more incomplete protein food sources, we can make a complete protein. For example, beans with rice is a complete protein. Pasta with cheese also makes a complete protein.

AMINO ACIDS

Amino acids are the building blocks of protein.

Protein builds, maintains, and replaces body tissue. Our muscles and organs are made mainly of protein. Our body can make eleven non-essential amino acids. But, it cannot make the other nine essential amino acids needed to build over two-thousand proteins. We have to get them from the food we eat.

FATS

Fats are macronutrients.

Fats are a source of energy in our food.

Monounsaturated fats, polyunsaturated fats, saturated fats and trans fats are four types of fats.

Fats belong to a group of substances called lipids. At room temperature, some fats are liquid and others are solid.

MONOUNSATURATED FATS

Monounsaturated fats are one of the unsaturated fats. This fat is healthy for our body and heart. It is called, "the good fat."

Olive oil and canola oil are high in monounsaturated fat. Seeds and avocados are also high in monounsaturated fat.

This kind of fat is liquid at room temperature.

POLYUNSATURATED FATS

Polyunsaturated fats are yet another kind of unsaturated fat.

It is a good fat.

Sunflower oil and corn oil are both polyunsaturated fats. Fats found in salmon and tuna are also polyunsaturated fats. Polyunsaturated fat is known as an essential fatty acid.

SATURATED FATS

Animal fat and some plants contain saturated fats. Saturated fat is bad for our health if we eat too much of it.

Meat, dairy, eggs, and butter are all sources of saturated fats.

Coconut oil, cotton seed oil, and palm kernel oil are saturated fats from plants. Saturated fats are solid at room temperature.

Donuts, cookies, and cakes are sometimes made with saturated fats. You should eat these foods only once in a while.

TRANS FATS

Trans fat is vegetable shortening. In the food production process, trans fats are made by adding hydrogen to vegetable oil, a process called hydrogenation. This process changes liquid oil to a solid.

Hydrogenated oils are bad for our health. They can cause heart disease.

TRANS FAT

Some chips, crackers, and cookies may contain trans fats.

Foods prepared with trans fats are unhealthy.

MICRONUTRIENTS

Our body needs micronutrients in small quantities to function properly. Vitamins and minerals are what we call micronutrients.

Vitamin C, A, D, E, K and the B-complex vitamins are all micronutrients.

The minerals fluoride, sodium, selenium, iodine, copper and zinc are also micronutrients. We get all of these nutrients through food.

VITAMIN C

Vitamin C is a water-soluble vitamin. That means our body cannot store it. We need vitamin C for tissue growth and repair. When we fall down and bruise our knee, vitamin C helps heal the bruise. When we eat oranges, broccoli, or cauliflower, we get vitamin C.

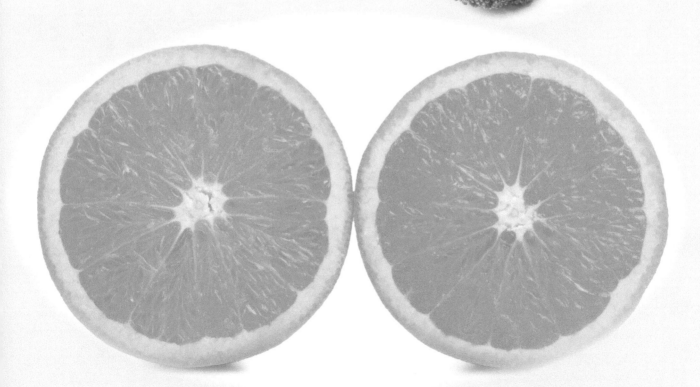

VITAMIN A

Vitamin A is a fat soluble vitamin. This means, vitamin A is stored in the fat tissues of our body for a few days or a few months. We need vitamin A to help keep our bones and teeth healthy. Foods that contain vitamin A are eggs, milk, cheese, and meat.

VITAMIN D

Vitamin D helps our body absorb and use calcium. It also helps form our teeth and bones. Vitamin D is a fat soluble vitamin.

When we eat tuna and eggs, we get vitamin D. When we drink milk, we can also get Vitamin D because this vitamin is often added to many kinds of milk.

VITAMIN E

Vitamin E is also a fat soluble vitamin.

It helps form red blood cells in our body. These cells deliver oxygen to tissues in our body.

It also aids in blood clotting. Clotting is when your body makes a scab after you cut yourself to keep you from bleeding too much.

Foods that contain vitamin E are whole grain products, avocados, and seeds.

VITAMIN K

Vitamin K is a fat soluble vitamin too.

It is also necessary for blood clotting.

Vitamin K also helps build and maintain strong bones.

Foods that contain vitamin K are chicken, cauliflower and eggs.

B-COMPLEX VITAMINS

B-Complex vitamins are a group of eight different vitamins. Each vitamin has its own name.

Thiamine, riboflavin, niacin, pantothenic acid, pyridoxine, biotin, folic acid, and cobalamin, form all the B-complex vitamins.

They are all water-soluble vitamins.

THIAMINE (B1)

Another name for the B-complex vitamin thiamine is B1.

Thiamine [thi-a-min] helps our body get energy from the food we eat. It turns the sugar in our blood into energy.

This process is called metabolism.

When we eat navy beans, kidney beans, and pork, we get the B-Complex vitamin thiamine.

RIBOFLAVIN (B2)

Riboflavin helps our body make red blood cells.

Another name for riboflavin is B2.

Yogurt, milk, and vegetables all contain riboflavin.

NIACIN (B3)

Another name for niacin [ni-a-sin] is B3.

Niacin, or B3, helps our body make energy.

It breaks down calories from protein, fats, and carbohydrates then turns them into energy.

Milk, meat, and whole grains are all great sources of niacin.

PANTOTHENIC ACID (B5)

Pantothenic [pan-to-then-ic] acid is also called vitamin B5.

Vitamin B5 helps our body digest and use fat.

When we eat carbohydrates, proteins, or fats, pantothenic acid turns them into energy.

PYRIDOXINE (B6)

Another name for pyridoxine [pyr-i-dox-ine] is vitamin B6.

Pyridoxine helps our brain function.

It also helps our body turn protein into energy.

When we eat bananas, oats and seeds, we get pyridoxine.

FOLIC ACID (B9)

Vitamin B9 is another name for folic acid.

When we eat foods that have folic acid, it helps our body make new cells, including red blood cells.

Oranges, chicken, and beans are excellent sources of folic acid.

COBALAMIN (B12)

Another name for cobalamin [co-bal-a-min] is vitamin B12.

Cobalamin, or vitamin B12, teams up with folic acid to help produce healthy red blood cells in our body.

Foods like eggs, chicken, and milk products, all contain cobalamin.

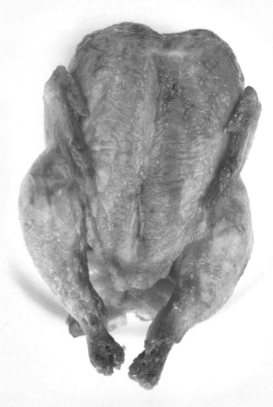

BIOTIN

Biotin is the eighth B-complex vitamin.

Biotin helps make fatty acids and glucose in our body from the food we eat. These substances provide our body with energy.

Cauliflower, carrots, and salmon all contain biotin.

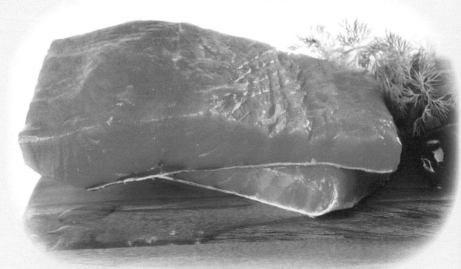

MINERALS

Macro minerals and trace minerals are two kinds of substances our body needs to stay healthy.

Macro minerals are the inorganic substances our body needs in larger amounts.

Trace minerals are the inorganic substance our body needs only in small amounts.

Minerals cannot be made in our body. We must get them from the food we eat.

MACRO MINERALS

Calcium, phosphorus, magnesium, sodium, potassium, chloride and sulfur are all macro minerals.

Our body needs these minerals in large amounts.

We get them through the food we eat.

CALCIUM

Calcium helps build strong bones and teeth.

Dairy products, broccoli, and canned salmon are good sources of calcium.

PHOSPHORUS

Phosphorus [Phos-pho-rus] helps make our teeth and bones.

Dairy products, seeds and lentils all contain phosphorus.

When we eat these foods, we get the macro mineral phosphorus.

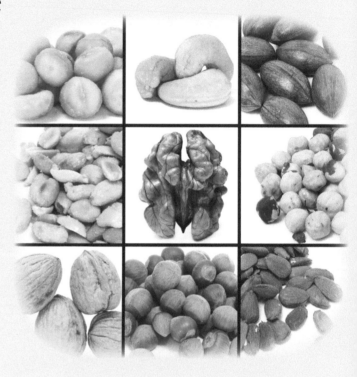

MAGNESIUM

Magnesium [mag-ni-si-um] controls the calcium levels in our body. It also helps maintain our skeletal structure.

Eggs, legumes, and vegetables are all dietary sources of magnesium. When we eat these foods, our body gets magnesium.

SODIUM

Sodium works with potassium.

It helps control the fluid levels in our body.

Another name for sodium is sodium chloride.

Canned vegetables and canned soups have lots of sodium. Many people often eat too much sodium. It is important not to eat too much sodium.

POTASSIUM

Potassium help control fluid levels in our body. It makes sure the amount of water in our body is just right.

Potassium also helps our muscles move.

Oranges, raisins, and potatoes are foods that contain potassium.

TRACE MINERALS

Trace minerals are essential for the health of our body.

Iron, manganese, copper, iodine, zinc, cobalt, fluoride and selenium are trace minerals our body needs only in tiny quantities.

IRON

Iron transports oxygen from our lungs to the rest of the body.

It helps form hemoglobin [he-mo-glo-bin].

Hemoglobin is a protein in red blood cells which carries oxygen through the body.

When we eat eggs, beans, and dried fruit, we get iron.

MANGANESE

Manganese [man-ga-nese] helps our bones grow.

When our bones grow, we get taller.

Manganese also helps our muscles work.

Beans, bananas, and strawberries all provide manganese.

COPPER

Copper, like iron, helps make hemoglobin in blood.

Copper also helps our body absorb and use iron.

Nuts, seeds, and green vegetables all contain copper.

IODINE

Iodine [i-o-dine] helps our thyroid work.

The thyroid is a gland located in our neck. It provides hormones and helps control the energy we use.

Milk, cheese, and fish all contain iodine.

ZINC

When we get a cut, zinc helps our cut heal.

Zinc also helps our bones grow.

Turkey, chicken, and whole wheat bread all contain zinc.

FLUORIDE

We need fluoride [fluor-ide] to help protect our teeth from cavities.

Fluoride also helps our bones stay strong.

Tap water has fluoride. And some kinds of bottled water contain fluoride too. When bottled water contains fluoride, it's called fluorinated.

SELENIUM

Selenium [se-le-ni-um] protects cells from damage causing free radicals. Free radicals are chemicals that cause cell damage in our body. When cells inside our body are damaged, they don't work as well as they should.

When we eat asparagus, eggs, and mushrooms, our body gets selenium.

WORDS YOU KNOW

Protein

WORDS YOU KNOW

CARBOHYDRATE

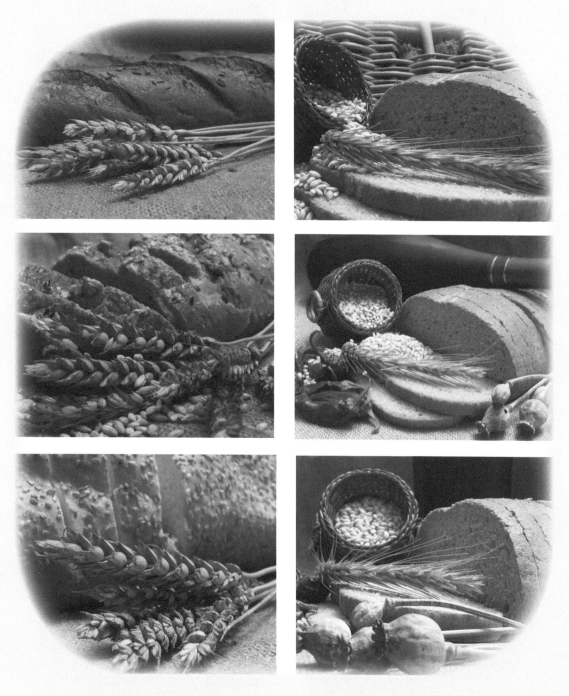

WORDS YOU KNOW

FAT

Organizations and Web Sites

MY PYRAMID FOR KIDS

www.mypyramid.gov/kids
The USDA Web Site on the new food pyramid.

NUTRITION EXPLORATIONS

www.nutritionexplorations.org/kids/nutrition-pyramid.asp
Explore the interactive food pyramid.

NATIONAL DAIRY COUNCIL

www.nationaldairycouncil.org/childnutrition
The Dairy Connection

PLACES TO VISIT

ARS National visitor Center
Powder Mill Road
Beltsville, MD 20705
301-504-9403
www.ars.usda.gov/is/nvc

Tour the visitor center for the USDA's Agricultural Research service, where you can learn about food science and nutrition.

Dietary Fat

Unsaturated Fat	Saturated Fat	Trans Fat
Food Examples	**Food Examples**	**Food Examples**
Almonds Vegetables Fish Olives Olive Oil	Beef Butter Pizza Ice Cream Lard	Cookies Donuts Cakes Fries Hydrogenated Oil
Benefits	**Benefits**	**Benefits**
Works in conjunction with saturated fats to prevent heart attacks and strokes Raises good cholesterol levels	Works in conjunction with unsaturated fats to prevent heart attacks and strokes	None

CPSIA information can be obtained
at www.ICGtesting.com
Printed in the USA
BVHW010101040323
659699BV00002B/7